Key Stage 2

Maths

Revision Notes

Author
Peter Patilla

Consultant
Mark Patmore

Letts
EDUCATIONAL

Every effort has been made to trace copyright holders and to obtain their permission for the use of copyright material. The authors and publishers will gladly receive information enabling them to rectify any error or omission in subsequent editions.

First published 1998
Reprinted 1998
Second edition 1999
Reprinted 1999
This edition 1999

Letts Educational, 9–15 Aldine Street, London W12 8AW
Tel. 0208 740 2270
Fax 0208 740 2280

Text © Peter Patilla 1998, 1999

Editorial, design and production by Hart McLeod, Cambridge based on books originated by Ken Vail Graphic Design

British Library Cataloguing-in-Publication Data

A CIP record for this book is available from the British Library

ISBN 1 84085 331 X

Printed in Italy.

Contents

Number and algebra

Measurement, notation and calculation

Shape, space and position

Handling data

These revision notes will help you to remember your work in the easiest way possible – by reducing the information you need to know to a series of brief facts and explanations. These will refresh your memory and help you understand the subject and organise your revision programme.

This book also contains key tips from examiners and space for you to write your own additional notes. Each section has a short test at the end so you can check you have covered each topic sufficiently.

This book will prove to be the key to success for your revision of Key Stage 2 maths. Revision of work and planning for exams are two vital steps on your learning path. This is a summary of the information you might find useful.

Start your preparation now. Ask your parents to help you plan the work using these notes.

In most subject areas you will have:
- coursework
- homework
- revision
- practice test papers
- a standard assessment test

The tests may make you feel worried and anxious, but with proper preparation, you will do well and might even enjoy them!

Make a list of what needs to be done, with your parents, and write down any coursework deadlines. Find out the dates of tests and arrange your revision timetable. Use the homework timetable to make a plan.

Divide your time between homework, coursework and learning. Ask your parents to support your work in the evenings or at weekends, but make sure plenty of time is included to relax, keep fit and have fun.

Revise several subjects each day – in 20-minute sessions.

Short bursts of revision in a variety of subjects are better than trying to revise one subject for a long time.

Get into the habit of revising at set times

Choose whenever is best for you to revise – early morning or in the evening. No one can concentrate well for long periods so make sure you take a break for about ten minutes between each 20-minute session. Twenty minutes of maximum concentration is better than an hour when your mind wanders.

Understand the work

Make sure you understand important concepts. If you have any difficulty with a topic, try explaining it to a relative or friend – this often helps the subject become clear. If you still don't understand, make sure you ask your teacher. Also, remember how important it is to use the correct words or terminology. The glossary at the back of the book will help you with some of these words.

Learn the work

Writing down or reading notes out loud often helps learning. Try to tell another person about the subject. With repetition, you will remember what you have learned.

Use practice questions

There is a limit to the number of questions examiners can ask. The more you practise, the less likely you are to be surprised. Remember, questions fall into three categories:
- knowledge – easily answered if you have learned the work
- understanding – easily answered if you have understood your work
- problem solving – testing your skill at using knowledge to interpret information

Get all the support and encouragement you can from family, friends and your school and enjoy learning.

Good luck with your revision and tests!

1 Number and algebra

PLACE VALUE AND THE NUMBER SYSTEM

Digits

There are ten digits which are: 0, 1, 2, 3, 4, 5, 6, 7, 8, 9.
These ten digits are used to build up all the other numbers we use.
2-digit numbers are the whole numbers from 10 to 99.

Examples
37 59 70 82 are 2-digit numbers.

3-digit numbers are the whole
numbers from 100 to 999.

Examples
104 368 742 are 3-digit numbers.

The word **digit** is often used to describe numbers.

Similarly there are numbers which have 4-digits, 5-digits, 6-digits and so on.

Place value

The column headings for the numbers we write are:

ten millions	millions	hundred thousands	ten thousands	thousands	hundreds	tens	units

The position of a digit in a
number changes its value.

A common error is to think that 'millions' comes immediately after thousands.

Examples

3 254	3 245	3 425	4 325
4 is worth 4	4 is worth 40	4 is worth 400	4 is worth 4000

It is important that you learn to say numbers in words
up to at least one million.

Examples
4022 ——→ four thousand and twenty-two.
36 295 ——→ thirty-six thousand, two hundred and ninety-five.
304 209 ——→ three hundred and four thousand, two hundred and nine.

Use these digits: **3, 7, 2, 1**
Arrange them to make the largest number and
the smallest number.

This question is a favourite type of test question.

Answer: largest **7 3 2 1**   smallest **1 2 3 7**

Multiplying by 10, 100 or 1000

To multiply by 10, move all the digits one place to the left.
The vacant place is filled in by a zero.

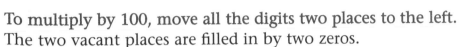

Examples
$3 \times 10 = 30$
$347 \times 10 = 3\,470$
$1204 \times 10 = 12\,040$
$3900 \times 10 = 39\,000$

Remember to fill the vacant places with zeros.

To multiply by 100, move all the digits two places to the left.
The two vacant places are filled in by two zeros.

Examples
$75 \times 100 = 7\,500$
$306 \times 100 = 30\,600$
$7520 \times 100 = 752\,000$

To multiply by 1000, move all the digits three places to the left.
The three vacant places are filled in by three zeros.

Examples
$24 \times 1000 = 24\,000$
$105 \times 1000 = 105\,000$
$3420 \times 1000 = 3\,420\,000$

Look out for this common error.

A common error is to say:
● add a nought when multiplying by 10
● add two noughts when multiplying by 100
● add three noughts when multiplying by 1000.

This only works with whole numbers: $34 \times 10 = 340$
It does not work with decimal numbers: $3.4 \times 10 = 34$ **not** 3.40

Dividing by 10, 100 or 1000

To divide by 10, move all the digits one place to the right.

Examples
$50 \div 10 = 5$
$470 \div 10 = 47$
$3700 \div 10 = 370$

To divide by 100, move all the digits two places to the right.

Examples
$300 \div 100 = 3$
$5800 \div 100 = 58$
$73\,400 \div 100 = 734$

To divide by 1000, move all the digits three places to the right.

Examples
$8000 \div 1000 = 8$
$13\,000 \div 1000 = 13$
$475\,000 \div 1000 = 475$

Ordering mixed whole numbers

When ordering mixed whole numbers it can help to write them underneath each other, lining up the unit column.

Example
Order 346, 56 074, 7204, 372 040

Write them in a list.
$$346$$
$$56\,074$$
$$7204$$
$$372\,040$$

It should now be easier to write the numbers in order.
Start with the smallest number first.

Comparing numbers

The symbols > and < are used to compare numbers.
> means greater than, < means less than.

Examples
15 > 8   6 < 11

The arrow shape of > and < points to the smaller number.

Rounding numbers

Rounding to the nearest 10:
● look at the units digit
● if it is less than 5, round down to the next 10
● if it is 5 or more, round up to the next 10.

| ≈ means is **approximately equal to.** |

Examples
34 ≈ 30   148 ≈ 150   375 ≈ 380

Rounding to the nearest 100:
● look at the tens and units digits
● if they are less than 50, round down to the next 100
● if they are 50 or more, round up to the next 100.

Examples
336 ≈ 300   764 ≈ 800   4550 ≈ 4600

Rounding up to the nearest 1000:
● look at the hundreds, tens and units digits
● if they are less than 500, round down to the next 1000
● if they are 500 or more, round up to the next 1000.

Examples
7436 ≈ 7000   8640 ≈ 9000   13 500 ≈ 14 000

Negative numbers

Some numbers are **positive** and some are **negative**.
Positive numbers are above zero and negative numbers are below zero.

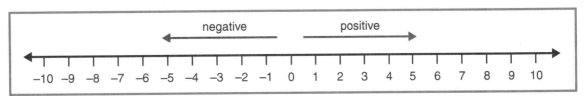

Example

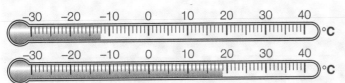

By how many degrees has the temperature risen?
Answer: 31°C

Example

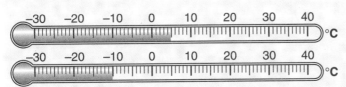

By how many degrees has the temperature fallen?
Answer: 15°C

Test questions often use negative numbers on thermometers.

Integer

An **integer** is any positive or negative whole number

... –3, – 2, –1, 0, 1, 2, 3, ...

Example
–2 is a negative integer.   2 is a positive integer.

Even numbers

Zero on its own is not an even number.
It represents nothing so cannot be odd or even.

- all **even** numbers are exactly divisible by 2
- all even numbers end in 0, 2, 4, 6 or 8.

Examples
8   36   410   3672   37 242

Odd numbers

- all odd numbers are not exactly divisible by 2
- all odd numbers end in 1, 3, 5, 7 or 9.

All odd numbers end in 1, 3, 5, 7 or 9.

Examples

7   63   411   7045   13 469

Square numbers

Numbers multiplied by themselves make **square numbers**.

Examples

$4 \times 4 = 16$

16 is a square number.

$7 \times 7 = 49$

49 is a square number

$12 \times 12 = 144$

144 is a square number.

The square numbers to 100 are: 1, 4, 9, 16, 25, 36, 49, 64, 81, 100

A short way of writing 3×3 is 3^2

Similarly $7 \times 7 = 7^2$ $15 \times 15 = 15^2$ $64 \times 64 = 64^2$

Square roots

Square roots are the opposites of square numbers.

To find the square root of a number you must find which number, when multiplied by itself, makes that number.

Number	1	4	9	16	25	36	49	64	81	100
Square root	1	2	3	4	5	6	7	8	9	10

Prime numbers

The only even prime number is 2.

Prime numbers are numbers which can only be divided by themselves and 1.

Here are some prime numbers:

2, 3, 5, 7, 11, 13, 17, 19, 23 ...

---

1 is never thought to be a prime number because it has only one factor – itself.

Multiples The multiples
of 2 are: 2, 4, 6, 8, 10, 12 and so on
of 3 are: 3, 6, 9, 12, 15 and so on
of 7 are: 7, 14, 21, 28, 35 and so on.

Multiples do not end at 10 times the number.

Notice that:
- multiples of 2 are all even numbers
- multiples of 5 end in 5 or 0
- multiples of 10 end in 0
- the digits of multiples of 9 always total 9, like this:

$$18 \rightarrow 1 + 8 = 9 \qquad 36 \rightarrow 3 + 6 = 9$$
$$27 \rightarrow 2 + 7 = 9 \qquad 45 \rightarrow 4 + 5 = 9$$

Multiples of a number do not end at ten times the number: they go on and on...
Some multiples of 5 are 45, 50, 55, 60, 65, 70 and so on.

Factors **Factors** are those numbers which will divide exactly into other numbers.

The factors of 12 are ⟶ 1, 2, 3, 4, 6 and 12
The factors of 15 are ⟶ 1, 3, 5 and 15.

They are sometimes organised into pairs like this:
The factors of 12 are   (1,12) (2,6) (3,4)
The factors of 15 are   (1,15) (3,5)
The factors of 100 are   (1,100) (2,50) (4,25) (5,20) (10,10).

RELATIONSHIPS BETWEEN NUMBERS

Number patterns and sequences

A **sequence** is a list of numbers.
There is usually a pattern to the numbers.

Example
3, 7, 11, 15 ... ➡   2, 4, 8, 16 ... ➡
The pattern, or rule, is add 4.   The pattern, or rule, is multiply by 2.

1, 3, 5, 7, 9...   odd numbers
2, 4, 6, 8, 10...   even numbers
1, 4, 9, 16, 25...   square numbers
1, 3, 6, 10, 15...   triangle numbers

Take note of these important sequences.

A special number sequence is
the **Fibonacci** sequence.
1, 1, 2, 3, 5, 8, 13 ... ➡
Each number is found by adding together the previous two numbers.

Equations

Equations have symbols or letters instead of numbers.
You need to work out what the symbol or letter stands for.

Example

$3 + \square = 15$ What number goes in the box?
Answer: 12

$4? + 2 = 50$ What number should replace the question mark?
Answer: 8

$x + 7 = 20$ What number does *x* stand for?
Answer: 13

Formulae

A **formula** (plural is formulae) uses letters or words.

A girl reads *p* pages of her book each day.
How many pages will she read in one week?
Answer: (7 × p) pages.

Four boys share *b* pounds equally.
How much will each receive?
Answer: £b or £b ÷ 4
        4

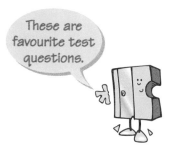

These are favourite test questions.

Function machines

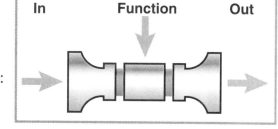

Function machines have three parts:
In, Out and the Function.

Examples

This function machine doubles and adds 1 to the number that goes in.

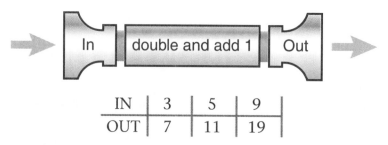

IN	3	5	9
OUT	7	11	19

Which numbers come out of each of these machines?

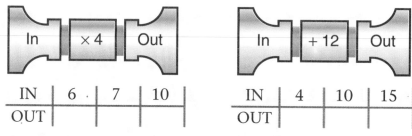

IN	6	7	10
OUT			

IN	4	10	15
OUT			

Answers: 24, 28, 40   *Answers: 16, 22, 27*

Sometimes you are asked which numbers go into the function machine.
You will need to know the **inverse**, or opposite, of the function on the machine.

- the opposite of adding is subtracting
- the opposite of multiplying is dividing.

Read these notes on inverses.

Which numbers go into each machine?

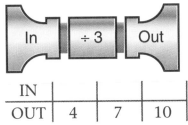

IN			
OUT	4	7	10

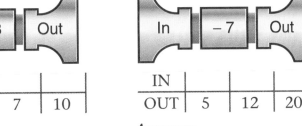

IN			
OUT	5	12	20

Answers:
(The inverse of ÷ 3 is × 3)
12, 21, 30

Answers:
(The inverse of – 7 is + 7)
12, 19, 27

Coordinates

Test questions often use shapes.

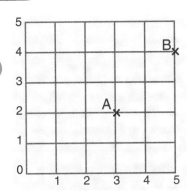

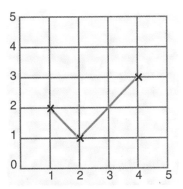

The coordinates of A are (3,2).
The coordinates of B are (5,4).
**The horizontal number is
always written first.**

Here are three corners of a rectangle:
(1,2) (2,1) (4,3).
What are the coordinates
of the fourth corner?
Answer: (3,4).

Coordinates also use negative numbers.

The coordinates of A are (2,3).
The coordinates of B are (3,–2).
The coordinates of C are (–3,–2).
The coordinates of D are (–2,3).

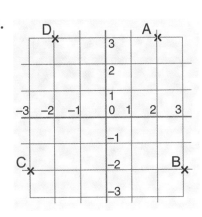

Addition

When adding several numbers, remember to write the units under each other.

Example

36 + 4694 + 245

$$
\begin{array}{r}
36 \\
4694 \\
+\ 245 \\
\hline
\end{array}
$$

When adding decimals, remember to put the decimal points under each other.

Example

4.4 + 0.37 + 14.2

$$
\begin{array}{r}
4.4 \\
0.37 \\
+\ 14.2 \\
\hline
\end{array}
$$

Add zeros to the 'empty' place if it helps.

Subtraction

When subtracting, remember the larger number goes at the top.

Example

4036 – 1987

$$
\begin{array}{r}
4036 \\
-\ 1987 \\
\hline
\end{array}
$$

When subtracting decimals, remember to put the decimal points under each other.
Make each number have the same number of digits after the decimal point by adding a zero, if necessary.

Example

3.4 – 1.74

$$
\begin{array}{r}
3.40 \\
-\ 1.74 \\
\hline
\end{array}
$$

Difference

Difference means how many more or how many less one number is than another.

Example

Find the difference between 56 and 80.
Answer: 80 – 56 = 24

Find the difference between £7.50 and £1.25.
Answer: £7.50 – £1.25 = £6.25

Multiplication

$$43$$
$$\times 25 \rightarrow$$

$$= (43 \times 20) + (43 \times 5)$$
$$= 860 + 215 = 1075$$

It doesn't matter which number you multiply by which.
Choose the way round you find easiest.

55×37 is the same as 37×55.

This is a useful tip.

Division

How to deal with remainders.

$$\begin{array}{r} 59 \quad r\,2 \\ 8\,\overline{)474} \end{array}$$ this is a **remainder** answer

$$\begin{array}{r} 59.25 \\ 8\,\overline{)474.00} \end{array}$$ this is a decimal answer

Sometimes you need to round up the remainder.

Example
Coaches can carry 50 children. How many coaches would be needed
to take 474 children to the seaside?
Answer: 9 coaches, remainder 24, so 10 coaches are needed.

Sometimes you need to round down the remainder.

Example
17 pennies are shared equally among 3 children.
How much will each receive?
Answer: 5p (the 2p over cannot be shared three ways).

Checking calculations

When checking calculations the process can be reversed like this:

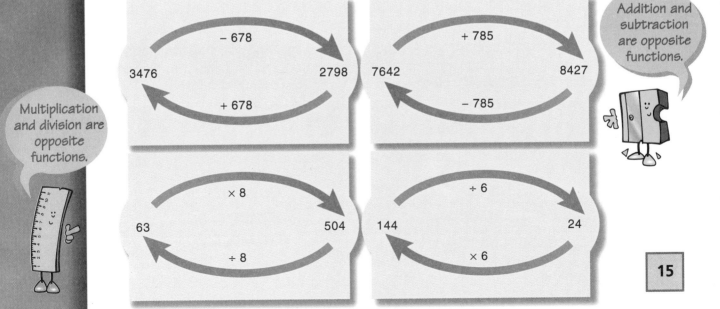

Addition and subtraction are opposite functions.

Multiplication and division are opposite functions.

15

Using brackets

Always work out the brackets first.

Example

$15 - (3 \times 4)$ $(15 \div 3) - 4$ $(12 + 3) - (20 - 6)$

$15 -\ \ 12\ \ = 3$ $\ \ 5\ \ - 4 = 1$ $15\ \ -\ \ 14\ \ \ = 1$

Mixed calculations

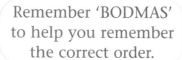

Sometimes there are several **operations**, such as adding and multiplying, in one sum. This is the order in which the sum should be worked out:

BRACKETS

OF

DIVISION

MULTIPLICATION

ADDITION

SUBTRACTION

Examples

$(246 \div 2) + 74 - 13$

$123 + 74 - 13$

$= 197 - 13$

$= 184$

Use the acronym BODMAS to help you remember the correct order.

Remember 'BODMAS' to help you remember the correct order.

$\frac{1}{2}$ of $16 + \frac{1}{3}$ of 12

$= 8 + 4$

$= 12$

Missing digits

This type of question really checks whether you know about calculations.

Examples

Work out one missing digit at a time.

These questions are favourite test questions.

```
   4 3 □          4 3 9          4 3 9
 +2 □ 4     →   +2 □ 4     →   +2 7 4
 ─────          ─────          ─────
  7 1 3          7 1 3          7 1 3
                    1              1
```

```
  9 1 □          9 1 0          9 1 0
 -□ 6 4     →   -□ 6 4     →   -3 6 4
 ─────          ─────          ─────
  5 4 6          5 4 6          5 4 6
```

It is important to check the finished sums are correct.

```
   4 □ 5          4 □ 5          4 6 5
 ×    □     →   ×    4     →   ×    4
 ───────        ───────        ───────
  1 8 6 0        1 8 6 0        1 8 6 0
```

```
   2 □ 5   →     2 2 5   →      2 2 5
 3)6 7 □        3)6 7 □        3)6 7 5
```

Addition and subtraction facts to 20

You should know all the addition and subtraction facts for numbers up to 20.

Learn addition and subtraction bonds by heart.

+	0	1	2	3	4	5	6	7	8	9	10
0	0	1	2	3	4	5	6	7	8	9	10
1	1	2	3	4	5	6	7	8	9	10	11
2	2	3	4	5	6	7	8	9	10	11	12
3	3	4	5	6	7	8	9	10	11	12	13
4	4	5	6	7	8	9	10	11	12	13	14
5	5	6	7	8	9	10	11	12	13	14	15
6	6	7	8	9	10	11	12	13	14	15	16
7	7	8	9	10	11	12	13	14	15	16	17
8	8	9	10	11	12	13	14	15	16	17	18
9	9	10	11	12	13	14	15	16	17	18	19
10	10	11	12	13	14	15	16	17	18	19	20

Learn multiplication and division bonds to 100.

You should know all your multiplication and division tables to 10 × 10 by heart.

×	0	1	2	3	4	5	6	7	8	9	10
0	0	0	0	0	0	0	0	0	0	0	0
1	0	1	2	3	4	5	6	7	8	9	10
2	0	2	4	6	8	10	12	14	16	18	20
3	0	3	6	9	12	15	18	21	24	27	30
4	0	4	8	12	16	20	24	28	32	36	40
5	0	5	10	15	20	25	30	35	40	45	50
6	0	6	12	18	24	30	36	42	48	54	60
7	0	7	14	21	28	35	42	49	56	63	70
8	0	8	16	24	32	40	48	56	64	72	80
9	0	9	18	27	36	45	54	63	72	81	90
10	0	10	20	30	40	50	60	70	80	90	100

What do you need to multiply to get all these numbers?
Use the tables on page 17 for help.

100									
90									
80	81								
70	72								
60	63	64							
50	54	56							
40	42	45	48	49					
30	32	35	36						
20	21	24	25	27	28				
10	12	14	15	16	18				
0	1	2	3	4	5	6	7	8	9

Doubles

You should be able to double quickly any whole number from 1 to 100, especially the tens numbers, such as 40, 60, 70.

Examples
Double 35 = 70   Double 30 = 60   Double 48 = 96

Halves

You should be able to halve quickly any whole number up to 20 and any tens number up to 100.

Halving an odd number gives a half in the answer.

Examples
Halve 18 = 9   Halve 17 = $8\frac{1}{2}$ Halve 50 = 25

Adding and subtracting 9 and 99

Use this method to add and subtract 9 and 99.

Example

36 + 9 ➡ add 10 then take away 1   *Answer: 45*

48 – 9 ➡ take away 10 then add 1   *Answer: 39*

248 + 99 ➡ add 100 then take away 1   *Answer: 347*

467 – 99 ➡ take away 100 then add 1   *Answer: 368*

Similarly, to add and subtract 19:

+ 19 ➡ add 20 then take away 1
– 19 ➡ take away 20 then add 1.

Using factors

When multiplying, you can break numbers up into their factors.

Examples

$8 \times 12 = 8 \times 3 \times 4$ or $8 \times 2 \times 6$ $12 \times 15 = 3 \times 4 \times 3 \times 5$

↑

break into factors   break into factors

When dividing, you can split the **divisor** up into factors.

Examples

$364 \div 14 = 364 \div$ (7 then by 2) ➡ $364 \div 7 = 52$

$52 \div 2 = 26$

Estimations and approximations

To check your answers, estimate like this:
- round the numbers up or down to easy numbers – usually tens or hundreds
- work out an estimate using the easy numbers
- use the symbol ≈ which means 'approximately equal to'.

Sometimes Ω is used instead of ≈.

Examples

$57 \times 34 \approx 60 \times 30 = 1800$

$726 \div 19 \approx 700 \div 20 = 35$

Your notes:

PLACE VALUE AND THE NUMBER SYSTEM

1 Write the value of the underlined digit.
  a) 3<u>4</u>7   **b)** 1<u>6</u> 729   **c)** 3 <u>4</u>27 965

2 Rearrange these digits ❹ ❻ ❶ ❾ to make:
  a) the largest number   **b)** the smallest number.

3 Multiply each number by 10.
  a) 56   **b)** 400   **c)** 5204

4 Divide each number by 100.
  a) 800   **b)** 56 000   **c)** 30 400

5 Write these measures in order of size.
Start with the smallest.
43 cm   4030 cm   304 cm   403 cm   3400 cm

6 Write each number to the nearest 10.
  a) 432   **b)** 875   **c)** 908

7 Write each number to the nearest 100.
  a) 750   **b)** 1149   **c)** 2781

8 Write each number to the nearest 1000.
  a) 3499   **b)** 17 801   **c)** 42 500

TYPES AND PROPERTIES OF NUMBERS

1 The temperature is –4°C.

a) If the temperature drops by 6°C what is the new temperature?

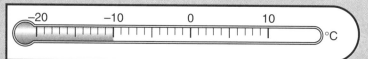

b) If the temperature rises by 12°C what is the new temperature?

2 Which of these are odd numbers?
87   72   135   250   363

3 Which of these are even numbers?
54   61   90   256   385

4 Which of these are multiples of 5?
58   85   90   65   70

5 Which of these are square numbers?

9   20   25   36   49

6 Write the square roots of these numbers.

a) 9   **b)** 25   **c)** 16   **d)** 36   **e)** 100

RELATIONSHIPS BETWEEN NUMBERS

1 Write the next two numbers in each pattern.

a) 15, 17, 19, 21, ..., ...   **b)** 42, 40, 38, 36, ..., ...

c) 7, 11, 15, 19, ..., ...

2 Write the missing numbers in these patterns.

a) 2, ..., 8, 16, 32, ...   **b)** 4, ..., 16, 25, 36 ...

c) ..., 30, 23, 16, ... 2

3 Solve these equations.

a) $\square + 17 = 23$ **b)** $17 - \square = 4$ **c)** $\dfrac{\square}{4} = 9$ **d)** $\square \times 4 = 24$

4 Solve these equations.

a) $x + 7 = 15$ **b)** $y - 6 = 12$ **c)** $\dfrac{p}{9} = 2$ **d)** $7q = 35$ **e)** $2m + 1 = 9$

5 a) Books cost £T each.   **b)** Three cakes cost £p.

What will 4 books cost?   What did one cake cost?

c) I think of a number, double it and take away 1. My answer is 7.

What number did I think of?

6 Complete the tables.

a)   **b)**

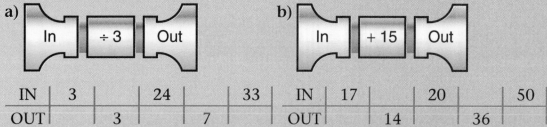

IN	3		24		33		IN	17		20		50
OUT		3		7			OUT		14		36	

7 a) Three corners of a square are: (3,0) (5,2) (3,4).

Write the coordinates of the fourth corner.

b) A line is drawn from (1,0) to (5,4).

Which of the coordinates (3,2) and (4,2) lie on the line?

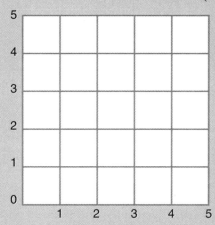

CALCULATIONS

1 Write the totals of:
  a) 436 + 78 + 6 045   **b)** 3.25 + 1.4 + 0.7
  c) John and Lisa each have 537 stickers. Susan has 294 stickers.
  How many have they altogether?

2 Work out the subtractions.
  a) 5020 − 1978   **b)** 3.2 − 1.78
  c) Leah has 3000 labels. She gives 1224 away. How many remain?

3 Find the difference between:
  a) 128 and 4000   **b)** 8.0 and 2.75
  c) Kim has 476 marbles and Alison has 614.
  How many more has Alison than Kim?

4 Work out the multiplication.
  a) 56 × 4   **b)** 236 × 7   **c)** 27 × 48
  d) Mr McDonald saves £17 each week.
  How much will he have saved at the end of the year?

5 Work out the divisions.
  a) 1 536 ÷ 6   **b)** 761 ÷ 5   **c)** 43.3 ÷ 4
  d) 8 sticks of rock are packed into boxes.
  How many boxes will be needed for 435 sticks of rock?

6 Work out the problems.
  a) 20 − (3 × 5)   **b)** (4 + 5) × (10 − 3)   **c)** (100 ÷ 4) + (100 − 36)

7 Find the missing digits.
  a)　　3 6 7   **b)**　　7 5 0   **c)**　　□□□   **d)**　1 3 8
  　　　+ □ 8 □　　　 − 3 6 □　　　× 　　4　　　3)□□□
  　　　　6 5 6　　　　 3 □ 2　　　　8 6 4

8 Write the answers.
  a) double 36   **b)** halve 70   **c)** 27 + 19   **d)** 285 − 99

2 Measurement, notation and calculation

TIME AND MEASURE

Equivalences

You need to know the following equivalent lengths and weights:

Mass and weight

Mass is closely related to weight but it is not the same. Mass is the amount of matter in an object. Weight is the force pulling the mass towards the centre of the Earth. Your weight is different on the Moon to that on Earth but your mass is the same.

1 metre	=	10 decimetres
		100 centimetres
		1000 millimetres
1 m	=	10 dm
		100 cm
		1000 mm
1 centimetre	=	10 millimetres
1 cm	=	10 mm
1 kilometre	=	1000 metres
1 km	=	1000 m
1 litre	=	10 decilitres
		100 centilitres
		1000 millilitres
1 l	=	10 dl
		100 cl
		1000 ml
1 kilogram	=	1000 grams
1 kg	=	1000 g
1 tonne	=	1000 kilograms
1 tonne	=	1000 kg

It is important to know what these prefixes mean.

deci = tenth part
centi = hundredth part
milli = thousandth part
kilo = 1000 times the unit

Example
0.75 litres = 7.5 dl
75 cl
750 ml

Some test questions might ask about the mass of an object.

Notice that there are different ways to write the same measurement.

How to convert between units.

Examples

2.5 cm	=	25 mm	2.5 litre	=	2500 ml	4.5 kg	=	4500 g
1.74 m	=	174 cm	1.75 litre	=	1750 ml	1.75 kg	=	1750 g
1.5 m	=	150 cm	0.125 litre	=	125 ml	3.125 kg	=	3125 g
1.25 km	=	1250 m						

$\frac{1}{2}$ cm = 5 mm

$\frac{1}{4}$ metre = 25 cm

$\frac{1}{2}$ metre = 50 cm

$\frac{3}{4}$ metre = 75 cm

$\frac{1}{4}$ litre = 250 ml

$\frac{1}{2}$ litre = 500 ml

$\frac{3}{4}$ litre = 750 ml

$\frac{1}{4}$ kg = 250 g

$\frac{1}{2}$ kg = 500 g

$\frac{3}{4}$ kg = 750 g

Remember simple fractions of measures.

Time

You need to know the following equivalent times.

1 minute = 60 seconds
1 min = 60 sec
1 hour = 60 minutes 3600 seconds 1 hr = 60 mins 3600 sec
1 day = 24 hours 24 hrs
1 week = 7 days
1 fortnight = 14 days
1 year = 12 months 365 days
leap year = 366 days

Months	Days
January March May July August October December	31 days
April June September November	30 days
February	28 days 29 days in a leap year

Telling the time

am are times in the morning
pm are times in the afternoon

midday = 12 o'clock = 12 pm
        = noon
midnight = 12 o'clock = 12 am

Example

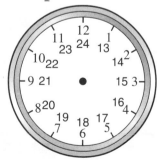

hours   minutes past
the hour

Learn about the 24-hour clock.

The 24-hour clock uses four digits.

Afternoon time											
in o'clock	1	2	3	4	5	6	7	8	9	10	11
24-hour clock	13	14	15	16	17	18	19	20	21	22	23

Morning times are between midnight and midday. A zero is used to fill the vacant space.

Example

9.45 in the morning   8 o'clock in the morning   3.10 in the morning

Afternoon times are between midday and midnight.

Example

12.30 in the afternoon   3 o'clock in the afternoon   10.20 in the evening

Notice how afternoon (and evening) o'clocks are 12 hours more than morning o'clocks on the 24-hour clock.

Examples

8 o'clock in the morning ➔ 08.00   10 o'clock in the morning ➔ 10.00
8 o'clock in the evening ➔ 20.00   10 o'clock in the evening ➔ 22.00

Sometimes two dots are used: Sometimes one dot is used:

Example   **Example**
09:30   13.25

How to write 24-hour times.

midnight   one minute after midnight   45 minutes after midnight

How to write midnight times.

Adding and subtracting measures

Always check you are adding and subtracting the same unit.

3 m + 175 cm + 45 cm   ➔ 300 cm + 175 cm + 45 cm = 520 cm
or   ➔ 3.00 m + 1.75 m + 0.45 m = 5.20 m
$\frac{1}{2}$ kg – 125 g   ➔ 500 g – 125 g = 375 g
$\frac{3}{4}$ litre + 0.75 litre + 200 ml   ➔ 750 ml + 750 ml + 200 ml = 1700 ml

Adding and subtracting time

Remember, when adding time, that
60 minutes make 1 hour.

$$\begin{array}{r} 2 \text{ hrs } 45 \text{ mins} \\ +1 \text{ hr } 55 \text{ mins} \\ \hline 4 \text{ hrs } 40 \text{ mins} \end{array}$$

Remember, when subtracting time, that
60 minutes make 1 hour.

5 hours – 1 hr 25 mins = 3 hours 35 minutes.

Be careful when
adding or
subtracting time.

Imperial units

Although we mainly use metric measures, some older units called
Imperial units are sometimes used.

Common
Imperial units
are still used.
Learn the rough
approximations
between metric and
Imperial units.

Length	Weight	Capacity
12 inches = 1 foot	16 ounces = 1 pound	8 pints = 1 gallon
$2\frac{1}{2}$ cm ≈ 1 inch	25g ≈ 1 ounce	1 litre ≈ $1\frac{3}{4}$ pints
30 cm ≈ 1 foot	$2\frac{1}{4}$ lb ≈ 1kg	$4\frac{1}{2}$ litres ≈ 1 gallon
Long distances use miles: 8 km ≈ 5 miles		

≈ means is **approximately equal to**.

PERIMETERS, AREAS AND VOLUMES

Perimeter   The distance around the outside edge of a shape.
Area   Amount of surface a 2-D shape covers.
   Units of area include cm², m² and km².
Volume   Amount of space a 3-D shape occupies.
   Units of volume include cm³ and m³.

Perimeters of shapes

Perimeters of
rectangles and
squares.

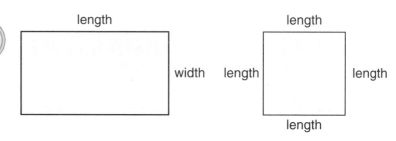

Perimeter = $l + w + l + w = 2l + 2w$ Perimeter = $l + l + l + l + = 4l$

The perimeter of this shape is: 4 + 2 + 6 + 6 + 10 + 8 = 36 cm.

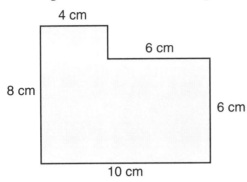

Sometimes you will need to calculate missing lengths.

Areas of irregular shapes

- count the whole squares
- squares which are $\frac{1}{2}$ or more count as whole ones
- ignore squares which are less than $\frac{1}{2}$.

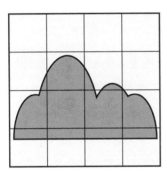

The area of the shape is approximately 5 squares.
Area ≈ 5 cm².

Areas of rectangles and squares

Area = length × width.
The most common units used are cm².

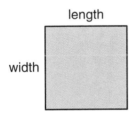

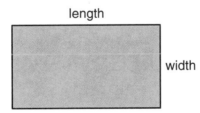

Example
area of 6 cm square: 6 × 6 = 36 cm².
Area of 3 cm by 8 cm rectangle: 3 × 8 = 24 cm².

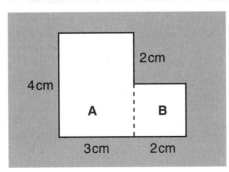

Split the area into two parts and find the area of each part.
Area A = 4 × 3 = 12 cm².
Area B = 2 × 2 = 4 cm².
Total area = 16 cm².

Calculate the area by adding.

27

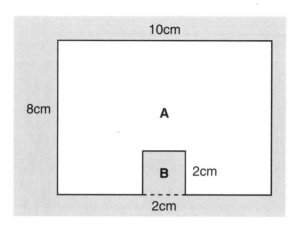

To find the total area, find the area of each part.
Area A = 8 × 10 = 80 cm².
Area B = 2 × 2 = 4 cm².
Total area = 80 – 4 = 76 cm².

Calculate the area by subtracting.

Areas of right-angled triangles

 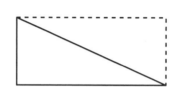

A square or rectangle can be drawn around a right-angled triangle.
The area is therefore half the square or rectangle.

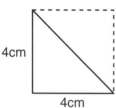 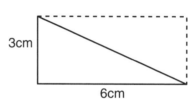

Area of the square = 4 × 4 = 16 cm².   Area of the rectangle = 3 × 6 = 18 cm².
Area of the triangle = $\frac{1}{2}$ × 16 = 8 cm².   Area of the triangle = $\frac{1}{2}$ × 18 = 9 cm².

Surface area

The surface area of a 3-D shape is found by totalling the area of each face.

- area of each end:
  2 lots of (3 × 5) = 30 cm²
- area of each side:
  2 lots of (3 × 4) = 24 cm²
- area of top and bottom:
  2 lots of (5 × 4) = 40 cm²
- total surface area = 94 cm².

Surface area of a cube:
6 lots of (4 × 4) = 96 cm².

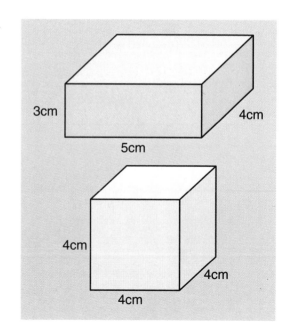

Circumference of circles

The circumference is the distance all around a circle.
The approximate circumference of a circle is 3 × diameter.

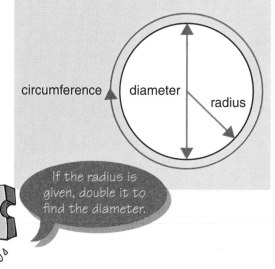

Example
The approximate circumference of a circle with a radius of 4 cm is:

diameter
↓
3 × 8 = 24 cm.

If the radius is given, double it to find the diameter.

To find the accurate circumference, we use the formula:
circumference = π × diameter
$c = \pi \times d$

The symbol π is read as 'pi'.

> π is a shorthand way of writing a long decimal number.
> π is approximately equal to 3.14.

Example
The circumference of a circle with diameter of 8 cm is: 3.14 × 8 = 25.12 cm.

Area of a circle

If the diameter is given, halve it to find the radius.

The approximate area of a circle is $3 \times r^2$.
r^2 is a short way of writing radius × radius.
r^2 = radius squared.

Example
The approximate area of a circle with a radius of 4 cm is: 3 × 4 × 4 = 48 cm².

To find the area accurately, use the formula:
Area = πr^2

Example
A circle with a radius 3 cm has an area of:
3.14 × 3 × 3 = 28.26 cm².

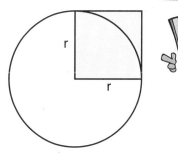

To find the area of a semi-circle, halve the area of the whole circle.

The area of a circle is πr^2.
The area of a semi-circle is $\dfrac{\pi r^2}{2}$.

How to calculate the area of a semi-circle.

29

Example

To find the area of a semi-circle with a radius of 5 cm:

$3.14 \times 5 \times 5 = 78.5$ cm^2.

The area of the semi-circle: $\frac{1}{2} \times 78 = 39.25$ cm^2.

Volumes of cubes and cuboids

The volume of a box can be found by counting how many centimetre cubes will fill the box.

Number of cubes in a layer   18
Number of layers   2
Volume   36 cm cubes

The volume of a 1 cm cube is written as 1 cm^3.
So we say that the volume of the box above = 36 cm^3.

Volume is given by: area of base × height = length × width × height.

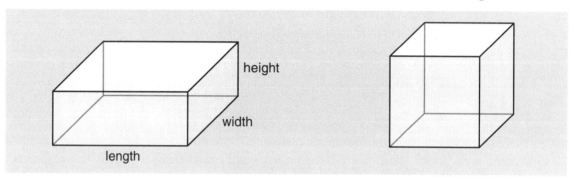

Example
Volume of a cuboid
2 cm × 3 cm × 5 cm = 30 cm^3.

Example
The volume of a 3 cm cube is $3 \times 3 \times 3 = 27$ cm^3.

Scale drawings

To make scale drawings of large objects, the lengths are all reduced by the same proportion.

Example
Here is the plan of a garden. Each centimetre on the plan represents 3 metres in the garden. We say the scale of the plan is: 1 cm to 3 m or 1 cm:3 m or 1:300 (the 3 metres has been changed to centimetres).

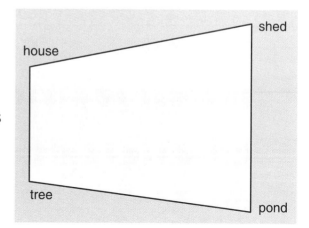

Maps use different scales:

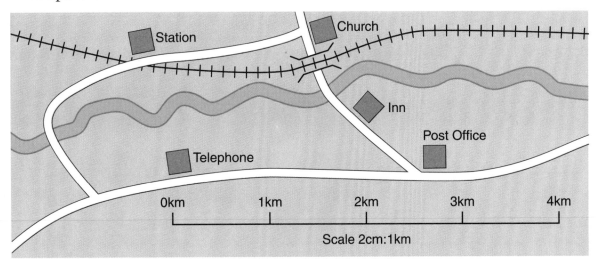

Example

How far is it from the inn to the telephone by road?

Answer: approximately 7cm = $3\frac{1}{2}$ km.

FRACTIONS, DECIMALS AND PERCENTAGES

Fraction names

A fraction is part of a whole number.

$\frac{2}{3}$ means 2 parts of 3.

The top number is the **numerator**.

The bottom number is the **denominator**.

$$\frac{2}{3} \quad \longleftarrow \text{numerator} \\ \longleftarrow \text{denominator}$$

A fraction like $\frac{3}{4}$ is called a proper fraction.

A fraction like $\frac{7}{4}$ is called an improper fraction because the top number is larger than the bottom number.

A fraction like $2\frac{1}{4}$ is called a mixed number because it contains both whole numbers and fractions.

Equivalent fractions

Equivalent fractions are worth the same but look different.

Example

$\frac{1}{2}$ is equivalent to $\frac{2}{4}$.

$\frac{2}{3}$ is equivalent to $\frac{4}{6}$.

31

Here are some equivalent fractions:

$\frac{1}{2}$ ➤ $\frac{2}{4}$ $\frac{3}{6}$ $\frac{4}{8}$ $\frac{5}{10}$ $\frac{6}{12}$ ... $\frac{1}{3}$ ➤ $\frac{2}{6}$ $\frac{3}{9}$ $\frac{4}{12}$ $\frac{5}{15}$ $\frac{6}{18}$ ... $\frac{1}{4}$ ➤ $\frac{2}{8}$ $\frac{3}{12}$ $\frac{4}{16}$ $\frac{5}{20}$ ...

To make equivalent fractions, multiply the numerator and denominator by the same number.

$$\frac{2 \times 4}{3 \times 4} = \frac{8}{12} \qquad \frac{3 \times 10}{4 \times 10} = \frac{30}{40} \qquad \frac{7 \times 2}{8 \times 2} = \frac{14}{16}$$

You can also divide the numerator and denominator by the same number.

$$\frac{12 \div 4}{16 \div 4} = \frac{3}{4} \qquad \frac{20 \div 5}{25 \div 5} = \frac{4}{5} \qquad \frac{8 \div 8}{16 \div 8} = \frac{1}{2}$$

Fraction of shapes

When dividing a shape into fractions the parts must be the same size.

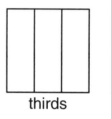

thirds   not thirds

It is usual to write the fraction in the simplest way, e.g. $\frac{1}{2}$ rather than $\frac{4}{8}$.

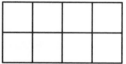

Fraction of quantities

To find the fraction of a quantity:
- divide by the denominator
- then multiply by the numerator.

Example
$\frac{1}{4}$ of £20 = £5 ➤ divide £20 by 4

$\frac{3}{4}$ of 24 cm ➤ find $\frac{1}{4}$ first

$\frac{1}{4}$ of 24 cm is 6 cm ➤ then find $\frac{3}{4}$ ➤ so $\frac{3}{4}$ of 24 is 3 × 6 = 18 cm.

Fractions on number lines

Fractions have positions on a number line.

Examples

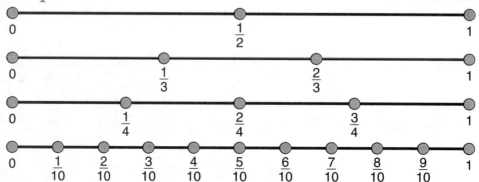

Mixed fractions

$1\frac{3}{4}$ is a mixed fraction.

$1\frac{3}{4} = \frac{7}{4}$ $\frac{23}{8} = 2\frac{7}{8}$

↖ 1 whole is four quarters.   ↖ 2 wholes are 16 eighths.

To change a mixed fraction into an improper fraction:
- change the whole number into its fraction, e.g. $4\frac{3}{4} = 4 = \frac{16}{4}$
- add this numerator to the numerator of the mixed fraction, e.g. $\frac{16}{4} + \frac{3}{4} = \frac{19}{4}$.

$\frac{11}{3} = 3\frac{2}{3}$ $\frac{57}{10} = 5\frac{7}{10}$ $\frac{16}{4} = 4$

3 thirds make a whole.   10 tenths make a whole.   4 quarters make a whole.

- note how many whole ones can be made
- make whole ones and the remainder is the numerator of the fraction.

Ordering fractions

When ordering fractions it can help to change all the fractions to the same denominator.

Example
To order:

$\frac{2}{3}$ $\frac{1}{2}$ $\frac{3}{4}$ $\frac{5}{6}$ ➡ change to twelfths:

It is now easier to order: $\frac{1}{2}$ $\frac{2}{3}$ $\frac{3}{4}$ $\frac{5}{6}$.

$\frac{2}{3}$	$\frac{1}{2}$	$\frac{3}{4}$	$\frac{5}{6}$
↓	↓	↓	↓
$\frac{8}{12}$	$\frac{6}{12}$	$\frac{9}{12}$	$\frac{10}{12}$

Decimals

A decimal point is used to separate whole numbers from fractions.

Example

tens	units		tenths	hundredths	thousandths
2	4	•	3	5	8
			↑	↑	↑
			3 means	5 means	8 means
			$\frac{3}{10}$	$\frac{5}{100}$	$\frac{8}{1000}$

Remember that hundredths are smaller than tenths.

Rounding decimals

Rounding to the nearest whole number:
- look at the tenths digit
- if it is less than 5 round down to the next whole number
- if it is 5 or more round up to the next whole number.

Remember ≈ means **approximately equal to**.

Examples

$4.3 ≈ 4$ $5.4 ≈ 5$ $3.5 ≈ 4$ $6.6 ≈ 7$

Rounding to the nearest tenth:
- look at the hundredths digit
- if it is less than 5 round down to the next tenth
- if it is 5 or more round up to the next tenth.

Examples

$0.4\underline{3} ≈ 0.4$ $2.5\underline{2}9 ≈ 2.5$ $3.2\underline{6}2 ≈ 3.3$ $5.3\underline{8}2 ≈ 5.4$

Multiplying decimals by 10 and 100

Multiplying by ten

Note how the decimal point moves one place to the right.
Any vacant spaces are filled with zeros.

Examples

$0.8 \times 10 = 8.0$
$4.9 \times 10 = 49.0$
$5.10 \times 10 = 51.0$
$13.11 \times 10 = 131.10$

Multiplying by 100

Note how the decimal point moves two places to the right.
Any vacant spaces are filled with zeros.

Example

$0.4 \times 100 = 40.0$
$5.2 \times 100 = 520.0$
$11.75 \times 100 = 1175.0$

Remeber these helpful facts

Decimals and fractions

It is important to know common fractions and their equivalencies.

$\frac{1}{2}$ is 0.50

$\frac{1}{4}$ is half this = 0.250

$\frac{1}{8}$ is half this = 0.125

Remember this tip!

$\frac{1}{10} = 0.1$

$\frac{1}{2} = \frac{5}{10} = 0.5$

$\frac{1}{4} = \frac{25}{100} = 0.25$

$\frac{1}{100} = 0.01$

$\frac{3}{4} = \frac{75}{100} = 0.75$

$\frac{1}{5} = \frac{2}{10} = 0.2$

Recurring decimals go on and on and on ...

$\frac{1}{3} \approx 0.3333$ This is a recurring decimal.

$\frac{1}{6} \approx 0.6666$ This is a recurring decimal.

3.4
3.40 } all are worth the same.
3.400

5
5.0 } all are worth the same.
5.00

Percentages

Percentages are fractions with a denominator of 100.

Example
$25\% = \frac{25}{100}$

Fractions can be changed to percentages by making the denominator 100.

$\frac{1}{2} = \frac{50}{100} = 50\%$ $\frac{1}{4} = \frac{25}{100} = 25\%$ $\frac{1}{10} = \frac{10}{100} = 10\%$

Finding percentages

The word **of** means **multiply**.
For example 20% **of** £200 means $\frac{20}{100} \times 200 = £40$.

Try to remember this important information.

Fractions, decimals and percentages

Fraction	$\frac{1}{2}$	$\frac{1}{4}$	$\frac{3}{4}$	$\frac{1}{5}$	$\frac{1}{10}$	$\frac{1}{8}$	$\frac{1}{3}$	$\frac{2}{3}$
Decimal	0.5	0.25	0.75	0.2	0.1	0.125	0.333	0.666
%	50%	25%	75%	20%	10%	$12\frac{1}{2}\%$	$33\frac{1}{3}\%$	$66\frac{2}{3}\%$

35

TEST

TIME AND MEASURE

1 a) Write 3.4 cm as millimetres.
  b) Write 2.125 litres as millilitres.
  c) How many kilogrammes in 4500 g?

2 a) How many days are there in a leap year?
  b) How many minutes are there in 3 hours?
  c) How many weeks are there in four years (including the leap year)?

3 Write the following times in the 24-hour clock:
  a) Half past six in the evening.
  b) Six o'clock in the morning.
  c) Midnight.
  d) Midday.
  e) Thirty-four minutes to nine in the evening.

4 Find the areas of these shapes.
  a)   **b)**   **c)**

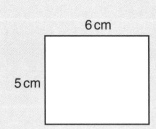

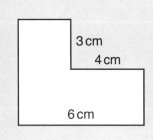

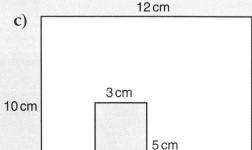

5 Add together these measures:
  a) 4 m + 50 cm + 5 mm
  b) 3 kg + 125 g
  c) 750 ml + 0.25 litres + 450 millilitres

6 Subtract these times:
  a) 6 hrs 45 mins – 3 hrs 40 mins
  b) 1 hr 30 mins – 45 mins
  c) 8 hrs 10 mins – 7 hrs 5 mins

7 Add these times:
  a) 4 hrs 20 mins + 4 hours 50 mins
  b) 7 hrs 45 mins + 1 hr 5 mins
  c) 9 hrs 10 mins + 6 hrs 25 mins + 3 hrs 45 mins

8 a) How many inches in 2 feet?
  b) How many ounces make approximately 50 g?
  c) About how many litres make 4 gallons?
  d) Approximately how many miles in 44 km?

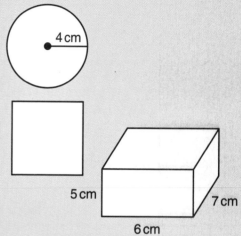

PERIMETERS, AREAS AND VOLUMES

1 Find:
   a) the circumference of the circle
   b) the area of the circle
      $\pi = 3.14$
      Circumference $= \pi \times d$
      Area $= \pi \times r^2$

2 The perimeter of a square is 36 cm.
   What is the length of one side?

3 a) Find the volume of this cuboid.
   b) Find the surface area of this cuboid.

4 Here is a scale drawing of a garden.
   Use a centimetre ruler to work out how wide the garden is in metres.

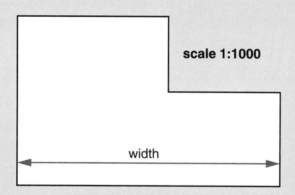

scale 1:1000

width

FRACTIONS, DECIMALS AND PERCENTAGES

1 Write the fraction shaded on each shape.
  a) 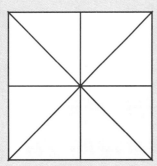   **b)** 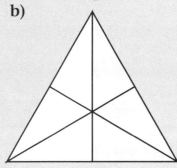   **c)**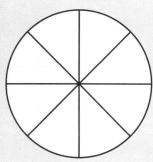

2 Write in the missing numbers.
  a) $\dfrac{1}{2} = \dfrac{}{8}$ **b)** $\dfrac{2}{3} = \dfrac{}{12}$ **c)** $\dfrac{7}{10} = \dfrac{}{100}$

3 Write the answers.
  a) $\dfrac{1}{2}$ of £16   **b)** $\dfrac{1}{3}$ of £21   **c)** $\dfrac{3}{4}$ of £8

4 Write the value of the underlined digits.
  a) 3.4̲6   **b)** 2.04̲   **c)** 15.2̲7

5 Multiply each number by 10.
  a) 4.6   **b)** 0.36   **c)** 22.30

6 Divide each number by 10.
  a) 3.4   **b)** 15.24   **c)** 0.7

7 Find 25 % of:
  a) £100   **b)** £40   **c)** £12

3 Shape, space and position

2-D SHAPES

Triangles

There are different types of triangles:

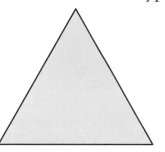

All shapes with three straight sides are called triangles.

equilateral
3 equal sides;
3 equal angles

isosceles
2 equal sides;
2 equal angles

scalene
no equal sides;
no equal angles

Sometimes triangles are described using angle words.

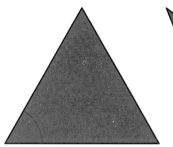

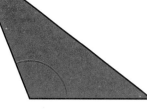

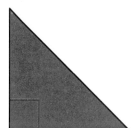

acute angled triangle
has angles less than 90°

obtuse angled triangle
has one angle more than 90°

right angled triangle
has an angle of 90°

Isosceles triangles right angled

can also be: obtuse

Scalene triangles can also be obtuse

Remember these types of triangles.

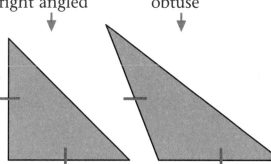

Quadrilaterals

All shapes with four straight sides are called quadrilaterals.

A square is a special rectangle. It is a rectangle with four equal sides and four right angles.

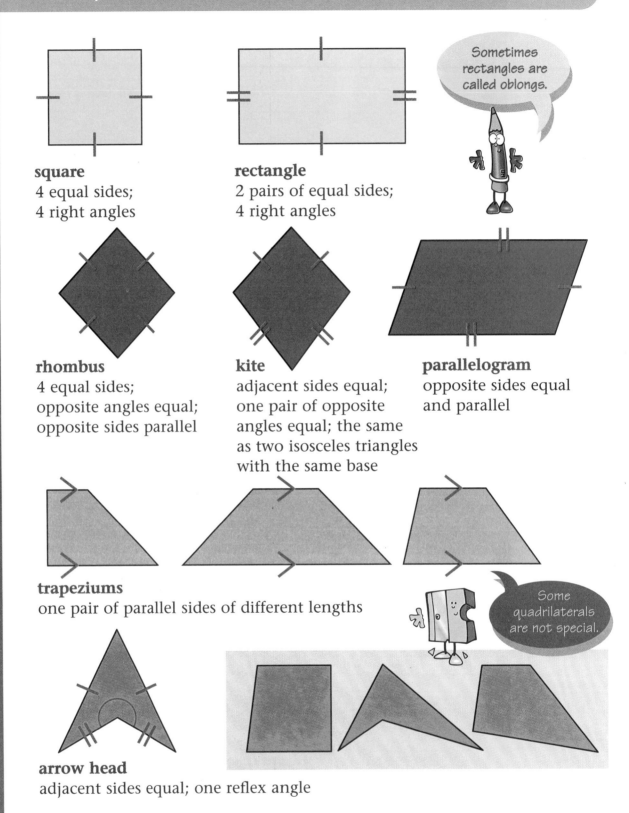

square
4 equal sides;
4 right angles

rectangle
2 pairs of equal sides;
4 right angles

Sometimes rectangles are called oblongs.

rhombus
4 equal sides;
opposite angles equal;
opposite sides parallel

kite
adjacent sides equal;
one pair of opposite
angles equal; the same
as two isosceles triangles
with the same base

parallelogram
opposite sides equal
and parallel

trapeziums
one pair of parallel sides of different lengths

Some quadrilaterals are not special.

arrow head
adjacent sides equal; one reflex angle

Hexagons Hexagons are any shape with six straight sides.

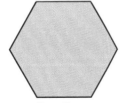

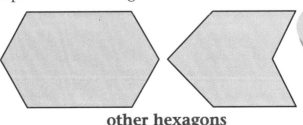

Remember, not all hexagons are regular.

regular hexagon
(has 6 equal sides)

other hexagons

Other 2-D shapes

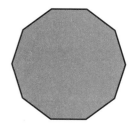

heptagon
7-sided shape

octagon
8-sided shape

nonagon
9-sided shape

decagon
10-sided shape

Regular shapes have sides that are all the same length and all their angles equal.
It is possible to have non-regular heptagons, octagons, nonagons and decagons.

Remember these facts.

Polygons The name of any shape with straight sides.

Number of sides	3	4	5	6	7	8	9	10
Name of polygon	triangle	quadrilateral	pentagon	hexagon	heptagon	octagon	nonagon	decagon
Regular shape								

Circles and ovals

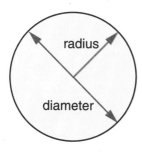

radius

diameter

circle
● circumference is all the way round

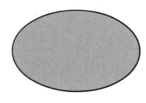

oval
● a flattened circle is also called an ellipse.

semi-circle
● a semi-circle is half a circle.

quadrant
● a quadrant is quarter of a circle..

Reflective symmetry

If a shape is symmetrical, then both sides are the same when a mirror line is drawn.

This is sometimes called mirror symmetry.

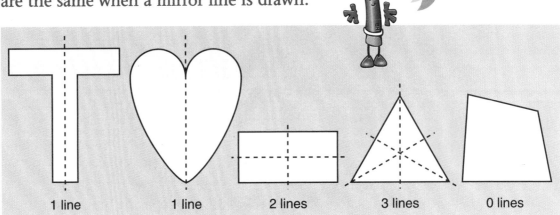

| 1 line | 1 line | 2 lines | 3 lines | 0 lines |

Sometimes you are asked to draw the reflection of a shape.

This is a favourite type of test question.

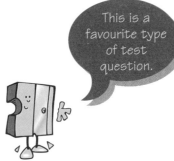

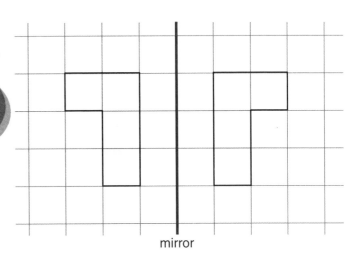

mirror

Rotational symmetry

When a turned shape looks exactly the same, it has rotational symmetry. All shapes have rotational symmetry, even if only to the order 1.

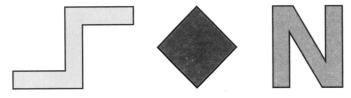

The order of rotational symmetry is the number of times the shape can turn and fit into its outline.

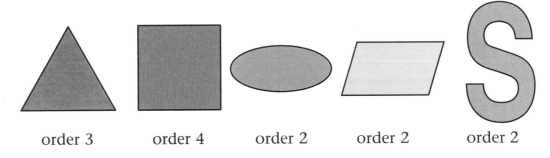

| order 3 | order 4 | order 2 | order 2 | order 2 |

Some shapes have both reflective and rotational symmetry.

 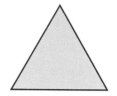 

Some shapes only have rotational symmetry.

 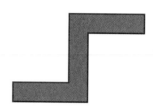

Congruent shapes

Congruent shapes are the same shape and size.
The shapes may have been flipped over.

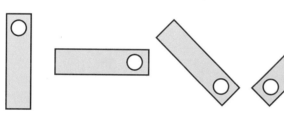

Congruent shapes
are identical in
shape and size.

Moving shapes

Moving by sliding
This is called **translation**.

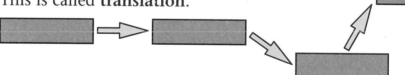

Moving by rotation
Rotation is either clockwise or anti-clockwise.

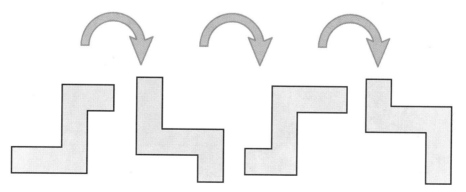

Clockwise rotation.

Moving by reflection

This is sometimes called flipping over.

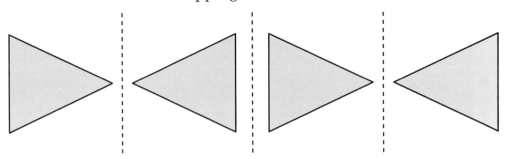

3-D SHAPES

Names of 3-D shapes

3-D shapes are sometimes called solid shapes.

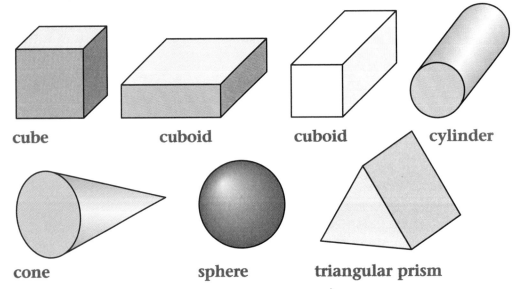

cube   cuboid   cuboid   cylinder

cone   sphere   triangular prism

> Prisms have two parallel faces and the same cross-section. The parallel faces can be any polygon.

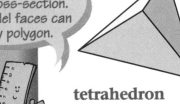
tetrahedron
(triangle-based
pyramid)

hemi-sphere

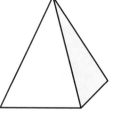
square-based
pyramid

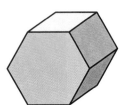
hexagonal
prism

a cube has:
8 corners (vertices)
12 edges
6 faces

> You need to know the parts of a 3-D shape.

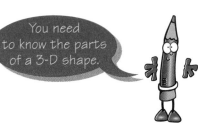

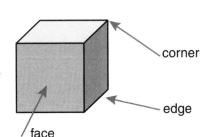
corner
edge
face

Polyhedron

The name of any 3–D shape which has polygon faces. Examples of polyhedron include prisms and pyramids. A cube is a polyhedron.

Number of faces	4	6	8	10	12	20
Name of polyhedron	tetrahedron	hexahedron	octahedron	decahedron	dodecahedron	icosahedron

Nets

The net of a shape is what it looks like when it is opened out flat.

net of a cube

net of a triangular prism

Some shapes have more than one net.

Some nets of cubes.

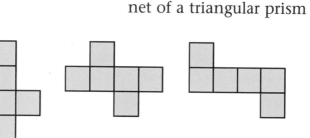

ANGLES

Names of angles

a right angle is 90°

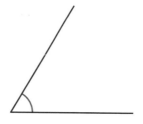

an acute angle is between 0° and 90°

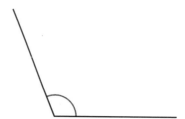

an obtuse angle is between 90° and 180°

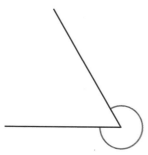

a reflex angle is between 180° and 360°

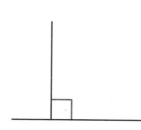

perpendicular lines meet at 90°

one rotation is 360° or 4 right angles

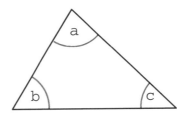
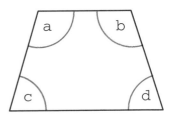

angles on a straight line add up to 180°

all the angles in a triangle add up to 180°

all the angles in a quadrilateral add up to 360°

$a + b = 180°$

$a + b + c = 180°$

$a + b + c + d = 360°$

Before measuring the size of an angle with a protractor, estimate its size. This will help you to read the correct size.

- Is it less than 90°?
  Yes.
- Is it more or less than 45°?
  More.
- Estimate the size.
  About 60°.
- Now measure the angle.

Estimating the size of an angle is an important skill.

Bearings A bearing is used instead of the points of a compass. It is always measured from north in a clockwise direction.

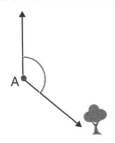

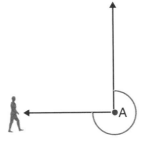

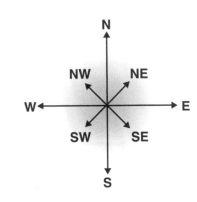

The bearing of the tree from A is 120°.

The bearing of the man from A is 270°.

Bearings of 90° or less start with a zero.
The bearing of the tree from A is 045°.

All bearings have three digits.

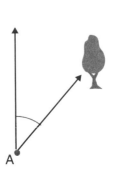

2-D AND 3-D SHAPES

1 Name each of these shapes.

a)   b) 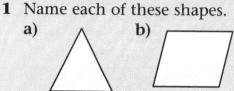 c) 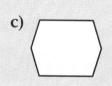 d)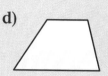

2 Name each of these shapes.

a) 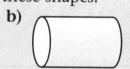 b) c) 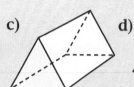 d)

3 Which of these shapes have rotational symmetry?

a)   b) 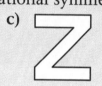 c) d)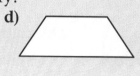

4 Which of these shapes is congruent to this?

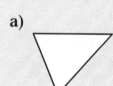

a) b)   c)   d)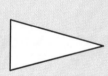

5 Which of these nets will make a triangular prism?

a) 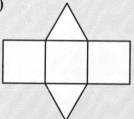 b) 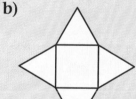 c)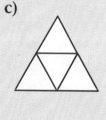

ANGLES

1 a) What is the size of angle A?
b) Is it acute or obtuse?

2 What is the bearing of tree B from tree A?

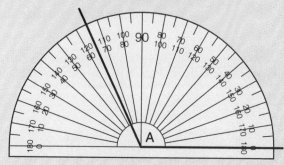

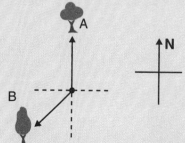

47

4 Handling data

GRAPHS

Pictogram

Small pictures are lined up to show the information.

Example
Sweets in a bag

= 2 sweets

There are 6 red, 5 green and 8 yellow sweets in the bag.

> Always check what scale is being used.

Bar graphs or column graphs

The columns can be horizontal or vertical.

Example

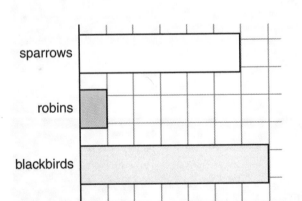

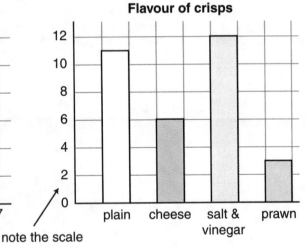

note the scale

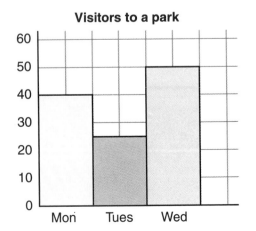

Sometimes the columns touch each other.

Stick graphs

Lines are sometimes used instead of columns or bars.

Example
Time taken for 50 swings of a pendulum of different lengths.

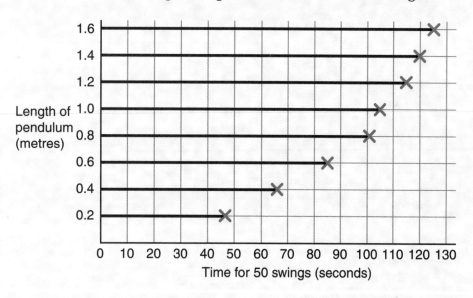

How many seconds did it take for 50 swings of a 0.6 m pendulum?
Answer: 85 seconds.

At what length did the pendulum take 100 seconds for 50 swings?
Answer: 0.8 m.

Line graphs

When the points are joined together it is called a line graph.

Example

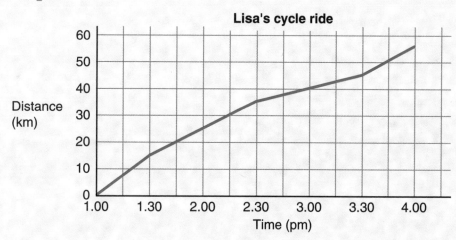

How far did Lisa cycle?   *Answer: 55 km.*

At what time had Lisa cycled 40 km?   *Answer: 3.00 p.m.*

Curved line graphs

Some line graphs are curved.

Example

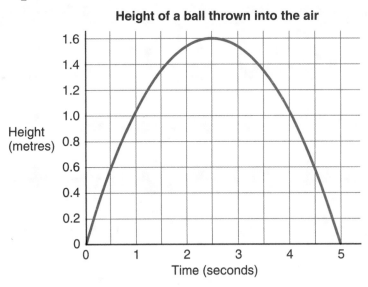

For how long was the ball travelling upwards?
Answer: $2\frac{1}{2}$ seconds.

What was the greatest height it reached?
Answer: 1.6 metres.

Conversion graphs

These are used to change one set of values to another.

Examples

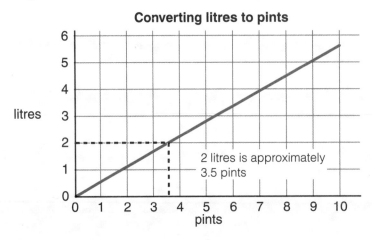

How many pints are approximately the same as 5 litres?
Answer: 8.8 pints.

How many litres are approximately the same as 9 pints?
Answer: 5.1 litres.

Arrow diagrams

Arrows can be used to show relationships.

Example

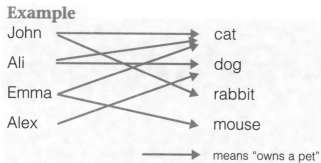

Who owns a rabbit?   *Answer: John.*

Who owns two pets?   *Answer: John, Ali and Emma.*

The arrow can be a number relationship.

Example

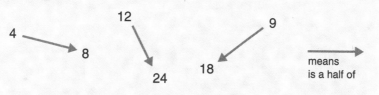

The arrow can be reversed.

What would the arrow mean if it were reversed?
Answer: "is double" or × 2.

Flow diagrams

Flow diagrams have instructions for you to follow.

Example

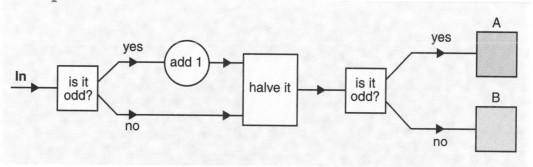

If 9 starts it ends up in Box A.
If 11 starts it ends up in Box B.
If 14 starts it ends up in Box A.
If 20 starts it ends up in Box B.

Venn diagrams

Venn diagrams are usually shown as overlapping circles or rings.
You must understand what each part of the diagram means.

Example

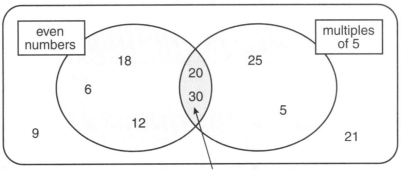

check you know what the overlapping region means

The area outside the rings is also important.

Which numbers are both even and multiples of 5?
Answer: 20, 30.

Write two numbers which are multiples of 5 and odd numbers.
Answer: 25, 5.

Write two numbers which are neither even numbers nor multiples of 5.
Answer: 9, 21.

Carroll diagrams

Carroll diagrams show two important aspects of each sorting such as:
odd and <u>not odd</u>, green and <u>not green</u>.

Example

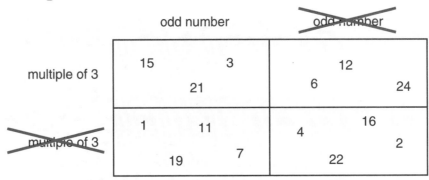

This is how 'is not odd' is usually written.

Which numbers are both odd numbers and multiples of 3?
Answer: 3, 15, 21

Write numbers that are odd but not multiples of 3.
Answer: 1, 7, 11, 19

Write numbers that are neither odd nor multiples of 3.
Answer: 2, 4, 16, 22

Tally charts

In a tally chart, one mark is made in a chart for each item.

Example

Favourite vegetables		Total
peas	ЖЖ III	8
carrots	ЖЖ I	6
cabbage	IIII	4

The fifth tally mark is put across the previous four: ЖЖ = 5.

Pie charts

These are circles divided into sections.
Each section represents a number of items.

Example
How much money did Sam save?
Answer: £1

How much money did Sam spend on magazines?
Answer: £1.50

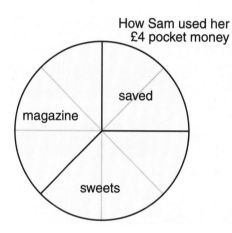

How Sam used her £4 pocket money

saved

magazine

sweets

Tables

Information is often shown in tables.

Work out what each column and row stands for.

Example
Popular boys' names.

	1944	1954	1964	1974	1984	1994
1st	John	David	David	Paul	Abdul	Thomas
2nd	David	John	Paul	Mark	James	James
3rd	Michael	Stephen	Andrew	David	David	Jack
4th	Peter	Michael	Mark	Andrew	Daniel	Daniel
5th	Warren	Peter	John	Richard	Michael	Matthew

Which was the top name twice in a row? *Answer: David*

In which year was Stephen in the top five names? *Answer: 1954*

Timetables

Timetabling questions are popular in tests.

Example

This is part of a timetable from Alfton to Emlock.

Alfton	10.20	10.46	11.20	11.53	12.16	12.46
Baxter	10.35	11.07	11.35	12.07	12.35	13.07
Coxmoor	10.42	11.15	11.42	12.15	12.42	13.15
Demly	10.50	11.23	11.50	12.23	12.50	13.23
Emlock	10.54	11.27	11.54	12.27	12.54	13.27

What time does the 11.20 train from Alfton arrive in Demly?
Answer: 11.50

How long does the 11.53 train from Alfton take to reach Emlock?
Answer: 34 minutes

AVERAGES

There are 3 types of average:

Mean $=$ $\dfrac{\text{total number of items}}{\text{number of items used.}}$

Median $=$ the middle value when the numbers are ranged in order of size.

Mode $=$ the number which occurs most often.

Range $=$ the spread of numbers (highest number – lowest number).

Mean

Examples
12, 5, 1, 4, 5, 5, 3

The mean of these numbers is $\longrightarrow$ $\dfrac{12 + 5 + 1 + 4 + 5 + 5 + 3}{7} = \dfrac{35}{7} = 5$

The range is $12 - 1 = 11$

Calculating the mean does not always produce a whole number.

Example
12 p   11 p   17 p   18 p   25 p $\blacktriangleright$ $\dfrac{83}{5} = 16.6\,\text{p}$

Some means can give strange answers.

To the nearest penny this is 17 p.

Median

Examples

To find the median of 3, 5, 2, 5, 11, 5, 4:
● put them in order 2, 3, 4, 5, 5, 5, 11
● circle the middle number 2, 3, 4, ⑤ 5, 5, 11.

The median is 5.
The range is 11 − 2 = 9.

> **Important:** If there are two numbers in the middle, the median is halfway between the two numbers.

Example

To find the median of 2, 4, 5, 3, 6, 3:
● put them in order 2, 3, 3, 4, 5, 6
● calculate the number halfway between the two middle numbers 2, 3, 3, ③.5, 4, 5, 6.

Mode

Examples

The mode of 3, 5, 5, 11, 5, 4 is 5 because it occurs most often.
The range is 11 − 3 = 8.

CHANCE AND PROBABILITY

> You need to know words which describe chance such as these...

Probability is the chance that something will happen.
Good chance, highly likely, certain, always, often, good likelihood, odds on, maybe, perhaps, occasionally, sometimes, evens...
No chance, unlikely, never, no likelihood, rarely...
Fair, unfair.

Probability scale

> This is called a probability scale.

All probabilities lie between 0 and 1.
You have to estimate the chance of any event happening.

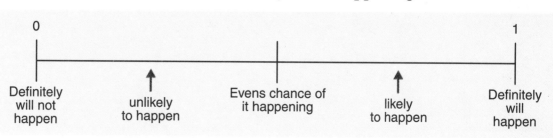

Example

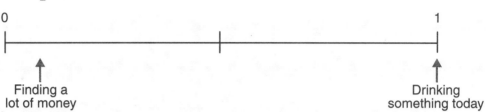

0   1

↑
Finding a
lot of money

↑
Drinking
something today

Calculating probability

Examples

Tossing a coin
There are two possibilities:
Heads or Tails
The chance of tossing heads is $\frac{1}{2}$.

Rolling a dice
There are 6 possibilities:
1, 2, 3, 4, 5 or 6
The chance of rolling each
number is $\frac{1}{6}$.

Spinners

A

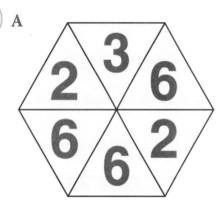

B

There are more 6s than any other
so there are more chances of
spinning 6 than the other numbers.

There are more 2s than any other
so there are more chances of spinning
2 than any other numbers.

On spinner A
What are the chances of getting 6?

*Answer: there are three 6s out of
six numbers.
The chance is $\frac{3}{6}$ or $\frac{1}{2}$.*

On spinner B
What are the chances of getting a 6?

*Answer: There are two 6s out of six
numbers.
The chance is $\frac{2}{6}$ or $\frac{1}{3}$.*

GRAPHS

1 People entering shops one morning.

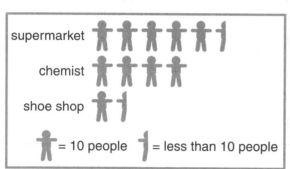

supermarket	
chemist	
shoe shop	

= 10 people   = less than 10 people

a) How many people visited the chemist?

b) How many people visited the shoe shop?

2 Graph of favourite pastimes.

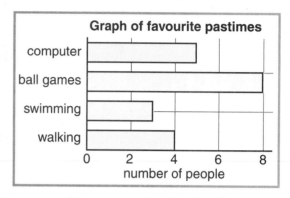

Graph of favourite pastimes

number of people

a) How many people played on a computer?

b) How many people were asked about their favourite pastimes?

3

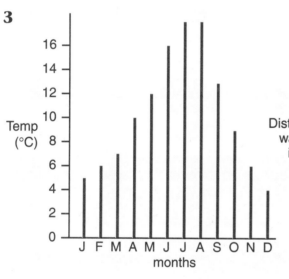

Temp (°C)

J F M A M J J A S O N D
months

a) Which month was the coldest?
b) Which month had a maximum temperature of 7°C?

4

Distance hiked by Lee Jones

Distance walked in km

10.00 11.00 12.00 noon 1.00 2.00 3.00
Time

a) How far did Lee walk?
b) Between what times did Lee rest?

5

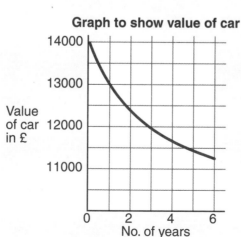

Graph to show value of car

Value of car in £

0 2 4 6
No. of years

a) What is the approximate value of the car after four years?

b) After how many years is the car worth £12 000?

6 a) How far from London was the train after four hours?
b) How long did it take the train to travel 150 km?

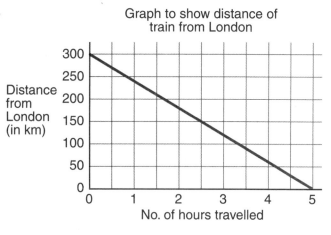

Graph to show distance of train from London

Distance from London (in km)

No. of hours travelled

DIAGRAMS

1 5 ———▶ 11   3 ———▶ 9   15 ◀——— 9
  6 ◀——— 0   10 ◀——— 4   10 ———▶ ☐

a) What do the arrows mean?
b) Which number goes in the ☐?

2

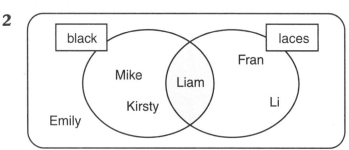

black   laces

Fran
Mike   Liam
Kirsty   Li
Emily

a) Who has black shoes with laces?
b) Who has lace-up shoes that are not black?

3 a) Who has a sister but no brother?
b) Who has no brothers or sisters?

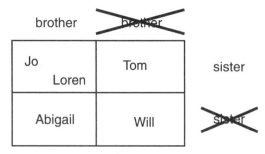

brother   ~~brother~~

Jo   Tom   sister
  Loren

Abigail   Will   ~~sister~~

4

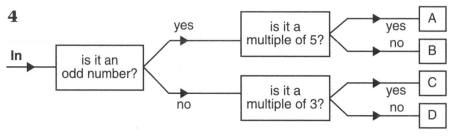

In ▶ is it an odd number? — yes ▶ is it a multiple of 5? — yes ▶ A
                                                          no ▶ B
            — no ▶ is it a multiple of 3? — yes ▶ C
                                            no ▶ D

Which box will these end up in?
a) 8   **b)** 30   **c)** 15   **d)** 100   **e)** 1

CHARTS AND TABLES

1 Visitors to Class 4

Monday	|||| |||| |||
Tuesday	|||| ||||
Wednesday	|||| |||| ||||	
Thursday	||||
Friday	||||

a) How many visitors were there on Monday?

b) Which day had the most visitors?

c) How many more visitors were there on Tuesday than on Friday?

2 Labels collected

	Mon	Tue	Wed	Thur	Fri
Sally	10	8	5	5	2
Thomas	3	5	0	7	6
Kapil	3	9	4	7	8
Lisa	5	6	9	9	2
Alison	0	6	2	3	6

a) Who collected most labels on Tuesday?

b) How many labels did Alison collect?

3 a) What time must you leave Leaton to arrive in Anminster at 13.50?

b) What time does the 17.00 train from Leaton arrive in Anminster?

Leaton depart	Anminster arrive
0700	0845
0800	0945
0900	1047
1000	1141
1100	1255
1200	1350
1300	1450
1400	1550
1500	1645
1600	1740
1700	1830

4 Visitors to the pantomime

girls

boys

men

women

60 girls visited the pantomime.

a) How many men visited?

b) How many people visited altogether?

59

AVERAGES

1 Find the mean of these amounts.
£22   £22   £32   £27   £42

2 What is the median of these numbers?
21   46   53   30   36   52   47

3 What is the mode of these shoe sizes?
7   8   7   8   9   8   7   8   9

4 What is the median and range of these numbers?
16   10   13   14   19   17

CHANCE AND PROBABILITY

1

Lisa's spinner
Jason's spinner

a) Who is most likely to spin a 3?
b) Who is most likely to spin an even number? (Think carefully!)

2

Wednesday will follow Tuesday

It will rain tomorrow

You will break a world record some time

Toast will land butter side down

a) Look at these events in order of the chance that they will occur. Start with the most likely.
b) Draw a probability line. Write the events in position on the line.

0   1

ANSWERS

 Number and algebra

Place value and the number system

1 a) 300 b) 6000 c) 400 000

2 a) 9641 b) 1469

3 a) 560 b) 4000 c) 52 040

4 a) 8 b) 560 c) 304

5 43 cm 304 cm 403 cm
   3400 cm 4030 cm

6 a) 430 b) 880 c) 910

7 a) 800 b) 1100 c) 2800

8 a) 3000 b) 18 000 c) 43 000

Types and properties of numbers

1 a) –10°C b) 2°C

2 (87 135 363)

3 (54 90 256)

4 (85 90 65 70)

5 (9 25 36 49)

6 a) 3 b) 5 c) 4 d) 6 e) 10

Relationships between numbers

1 a) 23, 25 b) 34, 32 c) 23, 27

2 a) 4 and 64 b) 9 and 49
   c) 37 and 9

3 a) 6 b) 13 c) 36 d) 6

4 a) $x = 8$ b) $y = 18$ c) $p = 18$
   d) $q = 5$ e) $m = 4$

5 a) £4T b) $\frac{£P}{3}$ c) 4

6 a)

3	9	24	21	23
1	3	8	7	11

b)

17	–1	20	21	50
32	14	35	36	65

7 a) (1,2) b) (3,2

Calculations

1 a) 6559 b) 5.35 c) 1368

2 a) 3042 b) 1.42 c) 1776

3 a) 3872 b) 5.25 c) 138

4 a) 224 b) 1652
   c) 1296 d) 884

5 a) 256 b) 152r1 or 152.2
   c) 10.83 d) 55

6 a) 5 b) 63 c) 89

7 a) 289 b) 368 382
   c) 216 d) 414

8 a) 72 b) 35 c) 46 d) 186

Measurement, notation and calculations

Time and measure

1 a) 34 mm b) 2125 ml c) 4.5 kg

2 a) 366 days b) 180 mins
   c) 208 weeks

3 a) 18.30 b) 06.00
   c) 00.00 d) 12.00 e) 20.26

4 a) 30 cm² b) 24 cm² c) 105 cm²

5 a) 4505 mm b) 3125 g c) 1450 ml

6 a) 3 hrs 5 mins b) 45 mins
   c) 1 hr 5 mins

7 a) 9 hrs 10 mins b) 8 hrs 50 mins
   c) 19 hrs 20 mins

8 a) 24 b) 2 c) 18 d) 28

Perimeters, areas and volumes

1 a) 25 cm² b) 50 cm²

2 9 cm

3 a) 210 cm³ b) 214 cm²

4 7 m

Fractions, decimals and percentages

1 **a)** $\frac{3}{8}$ **b)** $\frac{1}{2}$ **c)** $\frac{3}{4}$

2 **a)** $\frac{4}{8}$ **b)** $\frac{8}{12}$ **c)** $\frac{70}{100}$

3 **a)** £8   **b)** £7   **c)** £6

4 **a)** 4 tenths   **b)** 4 hundredths
    c) 2 tenths
    or $\frac{4}{10}$ or $\frac{4}{100}$ or $\frac{2}{10}$

5 **a)** 46   **b)** 3.6   **c)** 223 or 223.0

6 **a)** 0.34   **b)** 1.524   **c)** 0.07

7 **a)** £25   **b)** £10   **c)** £3

Shape, space and position

2-D and 3-D shapes

1 **a)** triangle   **b)** parallelogram
    c) hexagon   **d)** trapezium

2 **a)** cuboid   **b)** cylinder
    c) triangular prism
    d) tetrahedron
    or triangle-based prism
    or triangle-based pyramid

3 a and c have rotational symmetry

4 b

5 a

Angles

1 **a)** 115°   **b)** obtuse

2 225°

Handling data

Graphs

1 **a)** 40   **b)** between 10 and 20

2 **a)** 5   **b)** 20

3 **a)** December   **b)** March

4 **a)** 8 km   **b)** noon and 1.00 pm

5 **a)** £1160   **b)** 3 years

6 **a)** 60 km   **b)** $2\frac{1}{2}$ hours

Diagrams

1 **a)** +6   **b)** 16

2 **a)** Liam   **b)** Fran and Li

3 **a)** Tom   **b)** Will

4 **a)** D   **b)** C   **c)** A   **d)** D   **e)** B

Charts and tables

1 **a)** 13   **b)** Wednesday   **c)** 6

2 **a)** Kapil   **b)** 17

3 **a)** 12.00   **b)** 18.30

4 **a)** 90   **b)** 300

Averages

1 29

2 46

3 size 8

4 median 15   range 9

Chance and probability

1 **a)** Jason
    b) Jason and Lisa both have the
       same chance of $\frac{1}{2}$.

2 **a)** Wednesday will follow Tuesday
       Toast will land butter side down
       It will rain tomorrow
       You will break a world record
       sometime
    b)

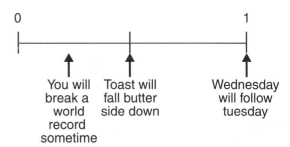

GLOSSARY

acute angle

an angle between 0° and 90°

bisector

a line which cuts another line or angle in two equal parts

chord

a straight line connecting two points of a curve

circumference

the length of a closed geometric curve, especially a circle

denominator

the bottom number of a fraction

diagonal

any line which joins two angles

diameter

a straight line connecting two points on the perimeter of a geometric figure

digits

the numerals 0 1 2 3 4 5 6 7 8 9

divisible

capable of being divided

divisor

a number or quantity to be divided into another number or quantity

horizontal

the horizontal is parallel to the horizon

improper fraction

a fraction with a bigger numerator than denominator

mixed number

whole number and fraction

multiple

a number which contains another an exact number of times

numerator

the top number of a fraction

obtuse angle

an angle between 90° and 180°

parallel

lines which remain the same distance apart

radius

a straight line which joins the edge of a geometric shape to the centre

reflex angle

an angle between 180° and 360°

vertical

the vertical is at right angles to the horizon

= equals sign   > greater than   < less than   ≈ approximately equals